Muhammad Abdul Rehman Kashif, Hafiz Tehzeeb Ul Hassan (Eds.)
Muhammad Salman Fakhar

Particle Swarm Optimization and its Variants

AF190747

Muhammad Abdul Rehman Kashif, Hafiz Tehzeeb
Ul Hassan (Eds.)
Muhammad Salman Fakhar

Particle Swarm Optimization and its Variants

For Non Cascaded Short Term Hydrothermal Scheduling

LAP LAMBERT Academic Publishing

Imprint

Any brand names and product names mentioned in this book are subject to trademark, brand or patent protection and are trademarks or registered trademarks of their respective holders. The use of brand names, product names, common names, trade names, product descriptions etc. even without a particular marking in this work is in no way to be construed to mean that such names may be regarded as unrestricted in respect of trademark and brand protection legislation and could thus be used by anyone.

Cover image: www.ingimage.com

Publisher:
LAP LAMBERT Academic Publishing
is a trademark of
Dodo Books Indian Ocean Ltd. and OmniScriptum S.R.L publishing group

120 High Road, East Finchley, London, N2 9ED, United Kingdom
Str. Armeneasca 28/1, office 1, Chisinau MD-2012, Republic of Moldova, Europe
Managing Directors: Ieva Konstantinova, Victoria Ursu
info@omniscriptum.com

Printed at: see last page
ISBN: 978-3-659-80066-5

Zugl. / Approved by: Lahore, University of Engineering and Technology, Lahore, 2014

Copyright © Muhammad Salman Fakhar
Copyright © 2015 Dodo Books Indian Ocean Ltd. and OmniScriptum S.R.L publishing group

Particle Swarm Optimization and Its Variants for Non Cascaded Short Term Hydrothermal Scheduling

Dedication

I dedicate this book to my beloved parents, my dear siblings and my niece Rumaisa.

Acknowledgment

I like to acknowledge first, Allah and His beloved Prophet (PBUH) as a gesture of thanks for without their will; it could have never been possible.

I also thank my parents and my siblings for providing me support throughout my education and for they assisted me whenever I felt down.

I also like to thank my teachers, the teachers of the Department of Electrical Engineering at U.E.T. Lahore. Special thanks is to my advisor Professor. Dr. Syed Abdul Rehman Kashif and to my teachers Professor. Hafiz. Tehzeeb Ul Hassan and Professor. Dr. Asghar Saqib, who showered there benevolences upon me throughout my studies.

Table of Contents

iv

List of Figures

List of Tables

Preface

In this book, a research work is presented, in which a famous short term hydrothermal scheduling problem is solved using a recently introduced variant of Particle Swarm Optimization algorithm. PSO and its variants fall in the category of meta-heuristic optimization techniques. By meta-heuristic, we mean randomness in the selection of particles. A comparative analysis among previously implemented algorithms, found in the literature, either non-heuristic or meta-heuristic, is given along with the implementation of Fully Informed Particle Swarm Optimization (FIPSO) algorithm on pumped storage hydrothermal scheduling problem.

The book chapters are divided and categorized in the following hierarchy:

Chapter 1 is all about the Introduction to the hydrothermal scheduling problem and its types like long term and short term hydrothermal scheduling. Moreover, it discusses briefly what types of works have already been done in this area of power system's optimization.

Chapter 2 discusses the algorithms of concern i.e. Fully Informed Particle Swarm Optimization and its parent form i.e. Canonical Particle Swarm Optimization. Both these algorithms are defined in detail and a 3 dimensional, multi-modal and non-linear optimization problem is shown to be implemented using the two techniques and a comparison is shown. Moreover, for the good practical understanding of the reader, two MATLAB program, one for each of the two algorithms is also provided.

Chapter 3 discusses the problem of interest of short term hydrothermal scheduling on which the PSO variants are supposed to be implemented.

Chapter 4 describes the implementation methodologies of the previously implemented PSO techniques on the problem described in chapter 3 of the book. Moreover, the implementation methodology of the FIPSO algorithm on pumped storage hydrothermal scheduling problem is also described.

Chapter 5 gives the results of the different implementations of PSO variants as found in the literature and also the results of the FIPSO algorithm when implemented on the pumped storage short term hydrothermal scheduling problem. After that, a final conclusion is given.

Abstract

The availability of electricity has become a major resource which describes the economic growth of any country in the modern world. In developing countries like Pakistan, extensive load shedding, due to the short fall of generated power, has motivated the power engineers and scientists to concentrate on the optimize use of the existing production of electrical power to alleviate the losses and the fuel consumption, in electric power generation, is reduced to minimize the cost of electricity produced. Moreover, the major portion of electrical power is generated through hydro-electrical systems in Pakistan which work in conjunction with the thermal power systems. While the cost of electricity produced though Hydro-electrical systems is quite low but the cost of electricity production through fossil fuels & natural gas is very high. Therefore, a hybrid of both these electrical generation process is utilized.

In this book, a comparative research work is done to solve a major power system optimization problem known as 'Short Term Hydrothermal Scheduling'. The technique utilized to solve this problem is Particle Swarm Optimization and its variants like Fully Informed Particle Swarm Optimization and accelerated PSO.

Chapter 1

Introduction

1 Overview

Electrical energy has become a basic utility in the modern world. Every society is working hard to enhance its production potential which requires more electrical power production to suffice not only the present demands but also enable to cater the future energy demands.

The concern to meet the future power demands does not only depend on a mere increase in power generating units but also it requires addressing the optimal operation of the existing power systems in most efficient and economic fashion. It is a well-known fact that hydroelectric power is one of the cheapest forms of electrical energy among all the modes of power generation. However, there are certain reasons for which the hydroelectric power cannot meet the demand on its own. In most of the countries, a hybrid of hydroelectric power and thermal power is used to meet the energy demands, which makes what we know as a hydro-thermal system. The major concern is to reduce the cost of production and related cost of any system and to increase the efficiency of the very system. Thus, it is desired to decrease the cost of operation and cost of production of hydro-thermal energy.

Short term hydro-thermal scheduling, one of the two types of hydro-thermal scheduling, has been implemented in some stochastic ways of optimization as well as some classical methods of optimization. It has always been the need to make the electricity production procedure even more cost effective. Particle Swarm Optimization (PSO) algorithm has recently been applied in many power system operation and control processes and is implemented in its canonical form as well as with some variations on the very problem of short term hydro-thermal scheduling. This book presents the works in the very domain taken from the literature, mentioned in references [1-20].

1.1 Hydro-thermal scheduling

Hydrothermal scheduling is a phenomenon in which the two systems (hydal and thermal) are dispatched in such a way so that the operation cost is minimized, which is mainly reflected by the fuel cost of the thermal units, by meeting the constraints of operation of both the hydro units and the thermal units [1]. Currently, most economic hydrothermal scheduling is a major concern in the power sector because of the mounting race in energy market [2].

There can be three types of hydro-electric power system.

In first type, there can be the power produced by only by hydro-electric systems i.e. there is no presence of any thermal unit. For such system, the production cost is not a major issue. However, the potentially available water is a major concern. Therefore, the constraints set for such a problem are concerned majorly with the forecasted and probably available water throughout the year. This type of problem is usually concerned with optimally best unit commitment rather economic dispatch because it requires mostly those units which can produce more energy by assuring minimized or optimized usage of water. However, the economic dispatch problem in this case is just to assure that the dispatching is so done that all the water releasing restraints are fully met.

There can be a system with a large number of hydro units and fewer numbers of thermal units. Now, the scheduling is done to meet the thermal constraints so that the fuel cost is reduced.

The third possible type can be the systems where the hydro units and thermal units are in equilibrium or the inclination is more towards the thermal units. Here, majorly, the thermal cost is a concern too, by meeting all the hydro restraints in parallel [3, 4].

3

1.2 Types of hydro-thermal scheduling

Depending upon the time, the hydro-thermal scheduling problem comprises of two major types.

1.2.1 Long-range hydro-thermal scheduling

Since hydro-thermal scheduling involves the requirement of water, therefore, the availability of water is assured. Depending upon the weather predictions, i.e. rainfall, snowfall, the scheduling of hydro units is carried out. In long-range hydro-thermal scheduling, the forecasts on the availability of water are made on the yearly basis. Depending upon these predictions, the hydro constraints are set and the hydro-thermal systems are scheduled. This type of problem is treated statistically and worst case scenarios are always kept in mind to set the constraints. The worst scenario can be the predictions regarding the yearly rainfall and snowfall which may get wrong and the availability of water does not suffice to meet the needs for the planned power production.

Such a problem of long-range scheduling can be performed by using various types of optimization techniques that range from conventional techniques like dynamic programming, Lag-range multiplier, and gradient search etc. to statistical production cost models to composite hydraulic simulation models and finally using meta-heuristic optimization techniques like genetic algorithms [3, 4].

1.2.2 Short-range hydro-thermal scheduling

Short term hydrothermal scheduling is short timed scheduling problem (one day to one week) and it is the hourly scheduling of all the generating units in such a way that such an amount of water be used so that the operating cost of the hydrothermal system be minimized. It addresses a type of problem in which the loading characteristics, water influxes and accessibilities are known most of the time. Starting circumstances like the volume of water in the reservoir are given and the constraints on the ending circumstances are also provided to fulfill the purpose of reducing the production and operation costs. It is also made sure to meet all the hydro and thermal constraints. The loss of power, which is majorly the function of hydro-power, usually first calculated and is presented as a function of hydro-power. These problems are usually non-linear, multi-dimensional and multi-modal. To solve these problems, an iterative procedure is mostly required and the problem is ended with optimally the best solution when the ending conditions are met. Usually the ending criteria is some sort of comparison of the variables' values with their previous iteration results or the ending criteria is the maximum number of iterations.

The hydrothermal scheduling problem can be defined as;

$$\min(F) = \sum_{j}^{N} n_j F_j \tag{1.1}$$

Subject to

$$\sum_{j}^{N} n_j D_j = D_{tot} \qquad \text{(Water discharge constraint)} \tag{1.2}$$

$$P_{load} + P_{loss} = P_{hydal} + P_{thermal} \qquad \text{(Load Balance)} \tag{1.3}$$

Where,

j = 1, 2, 3, 4... N   intervals

F_j → thermal generating cost

n_j → number of hours in j^{th} interval

P_{loss} is the transmission loss and is given by;

$$P_{loss} = f(P_{hydal})$$ (1.4)

Hydro generation is the function of discharge rate only;

$$P_{hydal} = f(D_j)$$ (1.5)

Where

$$\begin{cases} D_{min} < D_j < D_{max} & (Water\ discharge\ limits) \\ P_{thermal,min} < P_{thermal,j} < P_{thermal,max} & (Thermal\ generation\ limits) \\ P_{hydal,min} < P_{hydal,j} < P_{hydal,max} & (Hydro\ generation\ limits) \end{cases}$$

(1.6)

Also,

$$V_j = V_{j-1} + n_j(R_j - D_j - S_j)$$ (Reservoir volume at j+1 interval) (1.7)

Where,

'R_j' is water inflow rate at j^{th} interval

'S_j' is water spillage rate in the jth interval

$$V_{min} < V_j < V_{max}$$ (Reservoir storage limits)

It can be understood from the above statements that the main objective is to minimize the production cost of hydrothermal energy while meeting the hydro and thermal units' constraints.

1.3 Literature review

There are many methods to solve the short term hydrothermal problem. The conventional methods to solve the very problem are discussed in references [2, 4].

Stochastic optimization is a branch of mathematics and computer science which deals with the techniques with which optimal solution of an objective function is searched involving the randomness in a constructive way [5]. The very problem has been solved using some meta-heuristic optimization techniques also as discussed in references [1,2, 6-8].

Particle Swarm Optimization is a new optimization technique which is inspired from group behavior and the development of social norms. Among all the other stochastic optimization techniques, it is attaining a good popularity for its swift hurry towards convergence [9]. There are many modifications in the Canonical Particle Swarm Optimization algorithm as given above. In the further chapters, the problem of interest will be presented and its solution methodology using PSO variants found in the literature will be enlightened. In the final chapter, a comparison among different implementations as mentioned in [1] to [17] is also given.

Chapter 2

Canonical PSO and FIPSO

2 Overview

In this chapter, PSO in its canonical version and in the FIPSO version will be defined and explained. To elaborate further, a three dimensional multi-modal optimization problem will be solved using both the techniques, and the results will be compared.

2.1   Particle swarm optimization

PSO has now become a renowned meta-heuristic algorithm which was inspired from the social behaviors of the fish and birds while they search for food or find new habitats according to meet thesis needs. It was first presented by Kennedy and Eberhart in 1995. Though, the other famous meta-heuristic algorithms like genetic algorithm also follow the similar intelligence. Yet particle swarm optimization has gained fame for its ease in implementation for it uses the haphazardness of real numbers [10].

2.1.1 Canonical particle swarm optimization

In canonical particle swarm optimization algorithm, an objective function is optimized iteratively. This algorithm is stochastic in nature since at the first iteration, a set of possible solutions (particles) are randomly generated. Since these particles are the possible solution to the objective function, one out of these particles would be the best solution for that iteration. This best solution is known as the *global best solution* since it is globally providing the best solution to the objective function. Moreover, in the next iteration, each particle compares its own present value with its position of previous iteration. This gives the particle's local best. So, at the end of each of the iterations, every particle knows the following information;

- Global best particle '**P$_g$**'
- Particle's local best or its best position found so far '**P$_l$**'
- Particle's present position '**X$_i$**'

This information then sets the direction and magnitude of flow of particle **X$_i$** in the search space. Each particle **X$_i$** is influenced by the global best **P$_g$** and the local best **P$_l$** particles as shown in Figure 1 [10].

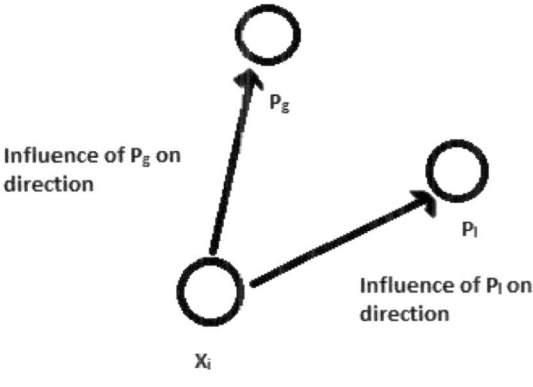

Figure 1: Influence of P$_g$ and P$_l$ on present value of particle in each iteration.

The canonical PSO iterations are proceeded as;

$$V_{i+1} = R\left(V_i + Rand\left(0,\frac{\varnothing}{2} \right).(P_i - X_i) + Rand\left(0,\frac{\varnothing}{2} \right).(P_g - X_i) \right)$$

(2.1)

$$X_{i+1} = X_i + V_{i+1}$$

(2.2)

Where

X_i $\rightarrow$ the existing position of the particle

X_{i+1} $\rightarrow$ new position of the particle

V_i $\rightarrow$ velocity to be added in previous location of particle to proceed forward.

R $\rightarrow$ restriction coefficient, approximately equal to 0.729 [11].

U $\rightarrow$ random vector generator function.

P_i $\rightarrow$ local best of particle.

P_g $\rightarrow$ global best of particles.

2.1.2 Fully informed particle swarm optimization

Since the birth of Particle Swarm Optimization algorithm, many variants of it have been coming into existence for it is believed that the canonical version is sometimes not able to reach the global optima of the objective function. A new extension to the canonical PSO was recently introduced by the author of the first version, i.e. Kennedy along with Mendes and Neves. It is observed that in classical PSO, each particle is not completely informed with its neighborhood. Therefore, reaching towards the optima is not that effective as it can be when fully informed [9]. However, in Fully Informed PSO, each individual particle is fully informed with its neighborhood. Its iterations are given as:

$$\vartheta_{i+1} = R \left(\vartheta_i + \frac{\sum\limits_{n=1}^{N_i} Rand(0,\phi).(P_{nbr(n)} - X_i)}{N_i} \right)$$

(2.3)

$$X_{i+1} = X_i + \vartheta_{i+1}$$

(2.4)

Where,

$P_{nbr(n)}$ → neighbor of particle Xi [11].

R → restriction coefficient approximately equal to 0.729.

ϑ_i → velocity of particle *i*.

By fully informed, it is meant that for each iteration, every particle X has the following information:

- Its own position X
- The information to the local best of each of its neighbor

Therefore, it is now possible for each of the particle to move the search space while getting an influence from all of the best possible locations found so far by each of its neighbor. And this is performed by every individual particle at the end of iteration as shown in Figure 2.

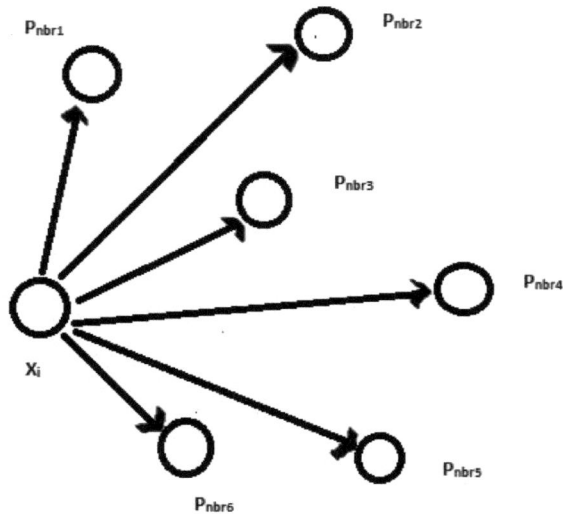

Figure 2: Influence on X_i of local best of each neighbor.

2.2 Neighborhood topologies in PSO

In Particle Swarm Optimization and the other Meta heuristic algorithms, though the heart is the generation of particles or the potential solutions randomly, however, a great deal of research is done in finding the performance of these algorithms by generating particle pseudo randomly. There are different types of graphs, specially generated to see the performance of the PSO algorithms and its variants. Though this type of pseudo randomness deviates from the originality of the primary algorithm yet this domain of research is such a hot topic that every other researcher is investing his/her time in it.

Following are the pseudo random or spatially predefined topologies of neighborhood implemented in PSO and its variants.

- **Gbest topology:** It's the one in which the particles are generated

13

randomly and the iterations proceed by getting influence of the global best among all the neighbors [12]. In this topology, all the particles of the array or particle matrix are considered to be the neighbors of each entrant of that array or matrix. One important thing to remember is that the particles are generated using uniform random number generator. There are many random number generation algorithms, however, it is usually preferred to use uniform random number generator function in the swarm intelligence algorithms [18].

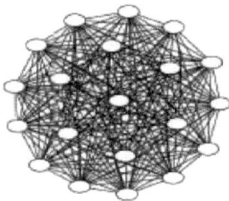

Figure 3: *Gbest* neighborhood of particles taken from [12].

- **Lbest topology:** in this topology, the particles are so generated that they form the shape of a ring and thus each particle has two immediate neighbors. As a result each particle is influenced by just two neighbors and iterations proceed by following the global better of the two neighbors per particle per iteration [12,18].

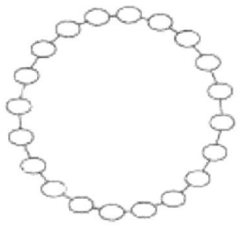

Figure 4: *Lbest* neighborhood topology taken from [12].

14

- **Pyramid topology:** In this topology a 3-D pyramid like particle's neighborhood is considered [12].
- **Mesh topology:** This topology is also known as Von Neuman topology of neighborhood. In this topology, the particles are arranged in the form of a grid of three dimensions [13].

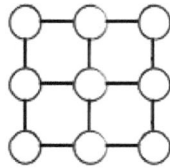

Figure 5: Mesh topology.

- **Star topology:** This topology is quite an unlike PSO topology. A central particle is the only one which moves on as an information handler while its neighbors just influences it. None of the other neighbors proceed as potential particles to be declared as global best [13].

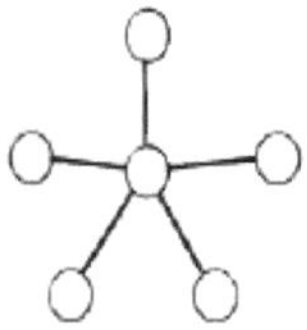

Figure 6: Star neighborhood topology taken from [13].

- **Toroidal topology:** this topology is not so different from the grid or mesh topology but its only difference is that it has just four neighbors to influence it [13].

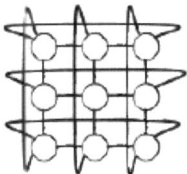

Figure 7: Toroidal neighborhood topology taken from [13].

- **Tree neighborhood topology:** In this topology, the particles are connected in the form of tree. It has a primary level node or particle followed by secondary and tertiary level root nodes of particles [13].

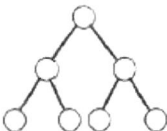

Figure 8: Tree neighborhood topology taken from [13].

2.2.1 Levels of connectivity and clustering in neighborhood

In all the above mentioned neighborhood topologies, the two major socio-metric factors that can vary the performance of each of the topologies are:

- *Level of connectivity* **k**
  - o It tells the number of neighbors a particle has.
  - o In *gbest* topology, k is equal to the total number of particles minus 1.
  - o In *lbest* topology, k is equal to 2.
- *Clustering* **C**
  - o It tells the number of neighbors common among different particles.
  - o This can be calculated on a single node/particle and for all the

16

nodes. The average of these calculations for each of the particle will give the value of C [12].

2.2.2 Impacts of different types of neighborhood on PSO

The research already made on these neighborhood topologies and their impact on the performance of different variants of PSO including the original PSO is quite interesting. It has been observed that for most of the functions, the *Gbest* topology does not work well, *Lbest* works better and the grid topology mostly outperforms all the other forms. It is worth mentioning that this comparison is made on the basis of statistical analysis relating to the convergence towards the optimum solution [12].

2.3 Solving an optimization problem using canonical PSO and FIPSO

To reveal and explain the Original Particle Swarm Optimization and the FIPSO algorithm, a test objective function is taken which is of two dimensions and is optimized by using the very algorithm. It is a multimodal function i.e. a function having numerous peaks. The function has been taken from source [10]. The 3 dimensional Michaelewicz function is given as

$$f(x, y) = -\left\{ \sin(x) \left[\sin\left(\frac{x^2}{\pi}\right) \right]^{2m} + \sin(y) \left[\sin\left(\frac{2y^2}{\pi}\right) \right]^{2m} \right\} \quad (2.5)$$

Where,

m = 10.

This two dimensional function has been solved using both the canonical PSO and FIPSO algorithms as given in reference [20]. The results are shown in the form

17

of table as well as in the diagram form.

The 3-dimensional Michaelewicz function is shown in Figure 9. It is well observed that this function is multimodal i.e. it has multiple number of minima. In such functions, the optimization algorithms often find a problem of sticking into the local minima of the locality in which the solution candidates or particles are moving.

It will be seen that the Particle Swarm Optimization algorithm will help in avoiding this problem. The mystery is solved by the random nature of the algorithm.

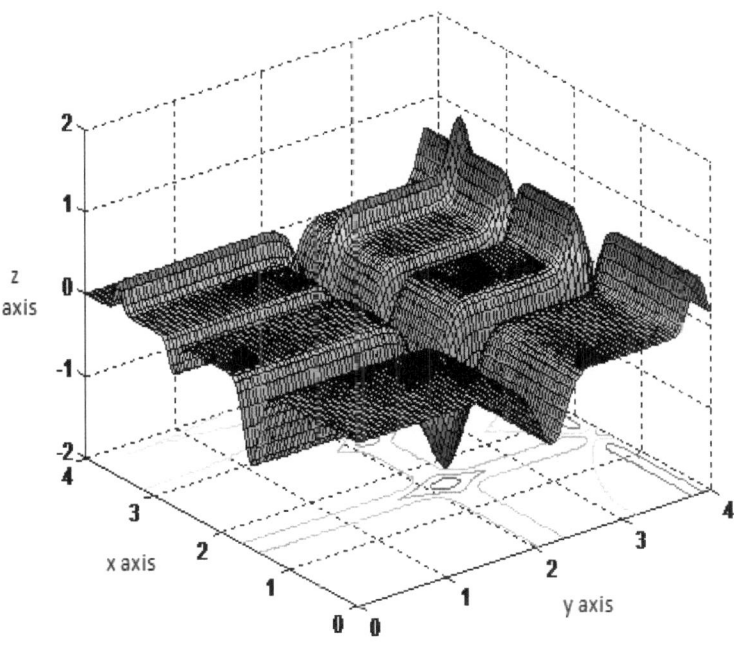

Figure 9: 3-D Michaelewicz function.

For the understanding of the readers, the MATLAB programs for this particular 3-D problem are given in the appendix.

2.3.1 What do the two results depict?

The two algorithms have been implemented and the results are shown in the form of tables which show the results of 15 iterations. In this case, the maximum number of iterations is taken to be the stopping criteria. Though the results are quite satisfactory and converging and there is not a great difference seen in the iterative process. However, the difference that actually took place was that fully informed PSO was able to span the complete search space for in it every particle is fully connected to every other particle for each of the iterations. However, it is not the case in canonical PSO. It is observed for this simple 3 dimensional problem that both PSO and FIPSO serve to converge to the global optimum. However, this is not the case every time, especially for the more complex functions, since both the algorithms are heuristic in nature. These programs have been run several times, and it is found that both PSO and FIPSO helped in reaching to the solution. One drawback is observed in the FIPSO algorithm for this simple 3D function is that it is unable to converge to the solution most of the times as discussed in reference [19]. And it is observed that for some spatially specified neighborhood topologies, like local best and mesh topologies, FIPSO works better than when global best neighborhood topology is considered as it is taken in the very research [9]. However, this impact of neighborhood topologies may also vary from problem to problem. This impact will be discussed in the chapter of results and the reader will be astonished to find that the observation regarding neighborhood topologies went totally opposite to as discussed in reference [9].

Table 1: Iterations of canonical PSO and best results of each iteration.

X	Y	Z= f (X,Y)
2.2096	3.3839	-0.7923
2.1654	2.4431	-0.7791
2.2556	3.1624	-0.7563
2.3728	1.5872	-1.4134
2.1115	1.5514	-1.6657
2.2687	1.6195	-1.6378
2.1088	1.6282	-1.5446
2.1721	1.6070	-1.7335
2.1817	1.5765	-1.7928
2.1927	1.5650	-1.7983
2.2009	1.5663	-1.8004
2.2418	1.5274	-1.7051
2.2768	1.5084	-1.5737
2.2179	1.5735	-1.7974
2.2179	1.5735	-1.7974

Table 2: Iterations of FIPSO and the best results of each iteration.

X	Y	Z= f (X, Y)
1.1537	1.6416	-0.8084
2.6956	1.5221	-0.9119
2.1515	0.8318	-0.7603
3.1796	1.5668	-0.9993
2.1515	1.3652	-0.9756
1.8848	1.6319	-0.9827
1.8739	1.6684	-0.7759
1.9312	1.5934	-1.1860
2.0142	1.5957	-1.3844
2.0881	1.5046	-1.4640
2.1064	1.5143	-1.5508
2.1865	1.5885	-1.7841
2.1617	1.5261	-1.6989
2.1757	1.5462	-1.7657
2.1824	1.5592	-1.7891

Chapter 3

Hydrothermal scheduling using FIPSO and other variants of PSO

3 Overview

In this chapter, two renowned hydro thermal scheduling problems i.e. non cascaded hydrothermal scheduling and pumped storage hydrothermal scheduling. A review of the methodologies to implement the former problem using different variants of PSO, as given in the literature, is presented. The later problem is solved using FIPSO so that the readers learn the way to implement the PSO algorithms on the two types of problem.

3.1 Non Cascaded Hydrothermal Scheduling Problem

A standard non cascaded short term hydrothermal scheduling problem is taken to practice upon.

A load is to be supplied from a hydro plant and a steam system whose characteristics are:

Corresponding thermal system:

$$
\begin{cases}
H = 500 + 8\left(P_{thermal}\right) + 0.0016\left(P_{thermal}\right)^2 & \text{(MBTU/hr)} \\
Fuel\ Cost = 1.15 \quad \text{(\$/MBTU)} \\
150\ MW < \left(P_{thermal}\right) < 1500 MW
\end{cases}
\tag{3.1}
$$

Hydro plant:

$$
D = \begin{cases}
330 + 4.97(P_{hydal})(\text{acre-ft/hr}) & 0MW \le P_{hydal} \le 1000 MW \\
5300 + 12\left(P_{hydal} - 1000\right) + 0.05\left(P_{hydal} - 1000\right)^2 & 1000 MW \le P_{hydal} \le 1100 MW
\end{cases}
\tag{3.2}
$$

Loading Outline:

Days	Hours	Power (MW)
1	First 12 hours	1200 MW
1	Second 12 hours	1500 MW
2	First 12 hours	1100 MW
2	Second 12 hours	1800 MW
3	First 12 hours	950 MW
3	Second 12 hours	1300 MW

Water-Reservoir Constraints are:

1. Volume of 100,000 acre-ft at the start

2. Volume of 60,000 acre-ft at the end of plan

3. Reservoir volume constraints are:

$$60,000(acre-ft) \leq V \leq 120,000(acre-ft)$$

4. There is a continuous inflow into the reservoir of 2000 acre-ft/h over the whole time schedule.

5. The continuity equation is given as

Volume $_j$ = Volume $_{j-1}$ + (Inflow $_j$ – Discharge $_j$ – Spillage $_j$) n $_j$ (3.3)

Here, the spillage is considered to be zero.

3.2 Problem solution using PSO and its variants

This problem has been solved using the conventional optimization techniques as well as the meta-heuristic optimization techniques as discussed in chapter 1 of the very thesis.

The very problem of hydro-thermal scheduling has been solved by Canonical PSO in two ways given below:

3.2.1 Hydro-thermal scheduling using PSO

This problem is firstly solved using canonical PSO by Samudai *et al.* [2]. In this research work, volume of water in the reservoir is taken as the particles and the other variables like discharge rate, thermal power and hydal power are taken as dependent variable. The main steps of this algorithm are given in [2] and represented in [19]:

1. Start haphazardly the reservoir volumes as particles for all the scheduling hours within the specified volume limits.
2. Get the global best for the 1st iteration and start the local best for each particle for the first iteration.
3. Generate randomly the velocity vectors between the minimum values to the maximum values.
4. Start the main iteration loop.
5. Find the global best for each of the iterations.
6. Find and upgrade the local best for each of the iterations.
7. For the next iterations, upgrade the particles' present locations.
8. For each of the iterations, evaluate the fitness function.
9. Stop the program.

The population size taken in the research is 50 particles and the maximum number of iterations for the program was taken to be 100.

3.2.2 Short term hydro-thermal scheduling using improved PSO

This research was conducted by Padamini *et al* [1]. In this research work rather than volume of water, water discharge rate was taken to be the best candidate for being a particle. However, the other three variables, i.e. hydro generation, volume of water in the reservoir and thermal generation were taken as dependent variables. The main steps of this research are mentioned in [1] and represented in [19]:

1. Generate randomly the water discharge rate as particles for all the scheduling hours within the specified discharge rate restrictions.
2. Find the global best for the first iteration and generate the local best.
3. Start randomly the velocity vectors between the lowest values to the highest values.
4. Start the main iteration loop.
5. Find the global best for each of the iterations.
6. Find and update the local best for each of the iterations.
7. Update iteratively the particles' locations.
8. For each of the iterations, evaluate the objective function.
9. Stop the program and print the results when the stopping criteria are reached.

In this research work, the number of population size was taken to be 30 particles and the maximum number of iterations was taken to be 100.

3.3 Pumped Storage Hydrothermal Scheduling Problem

A standard pumped storage short term hydrothermal scheduling problem is taken to practice upon, from reference [4]. It is a pumped storage hydrothermal scheduling problem in which during the off peak loading hours, the hydal plant is of pumped type, i.e. the reservoir is being filled back during these hours. The interested readers can refer to reference [4], example 7D, for understanding the type of problem. Given below is the implementation of FIPSO, a variant of PSO on the very problem.

3.3.1 Hydro-thermal scheduling using FIPSO

Fully Informed Particle Swarm Optimization is the variant of interest in this topic. This variant is not a very old entry in the domain of particle swarm intelligence and was introduced by the author of the canonical PSO himself along with Mendes [9].

This algorithm is taken to implement on the problem of interest for:

1. It allows the particles to set their new directions or velocities by getting influenced from all the particles in its neighborhood.
2. This full information allows each particle to span the complete search space and thus chances to reach the global maximum at the end of program obviously enhance which sometimes is not possible using canonical PSO for in it the particle is influenced only by the global best and its own local best information found.
3. Though the influence of all neighbors enhances the time but this increase in time is compensated by the fact that the number of neighbors required are reduced which lessens the time required. So, time doesn't remain a major issue.

In this research work, volume of water is taken to be the particle of interest as it helps in scanning the complete search space. Moreover, the reason behind choosing volume of water in the reservoir to be the particle is that its constraints are primarily given in the problem of interest as is described in the third chapter of the thesis.

The main points or flow of moving the problem are described below as taken from reference [18]. Interested readers are encouraged to read reference [18] for detailed discussions.

1. Generate randomly the particles arrays. For this random number generator function is used.
2. Initialize randomly the velocity two dimensional array of dimension 6 by 8. Velocity is defined within the lowest and highest limits.

$$
\begin{cases}
V_{max} = \dfrac{X_{max} - X_{min}}{no.\ of\ iterations} \\
V_{min} = -(V_{max})
\end{cases}
\tag{3.4}
$$

3. Initialize randomly the array for local bests of dimension 6 by 8.
4. Calculate arrays of hydro-power, thermal power, discharge rate, individual cost and minimum cost.
5. Make sure that the thermal and hydal power limits are not violated.
6. Calculate the objective function update the arrays of local bests.
7. Apply FIPSO main equations to upgrade the position and velocity arrays of the particles.
8. Repeat the iterations till the iteration number is reached to end.
9. Print the results.
10. Draw the graph between cost and iteration number to get the convergence behavior.

The above steps are shown in the form of a flow chart in Figure 10.

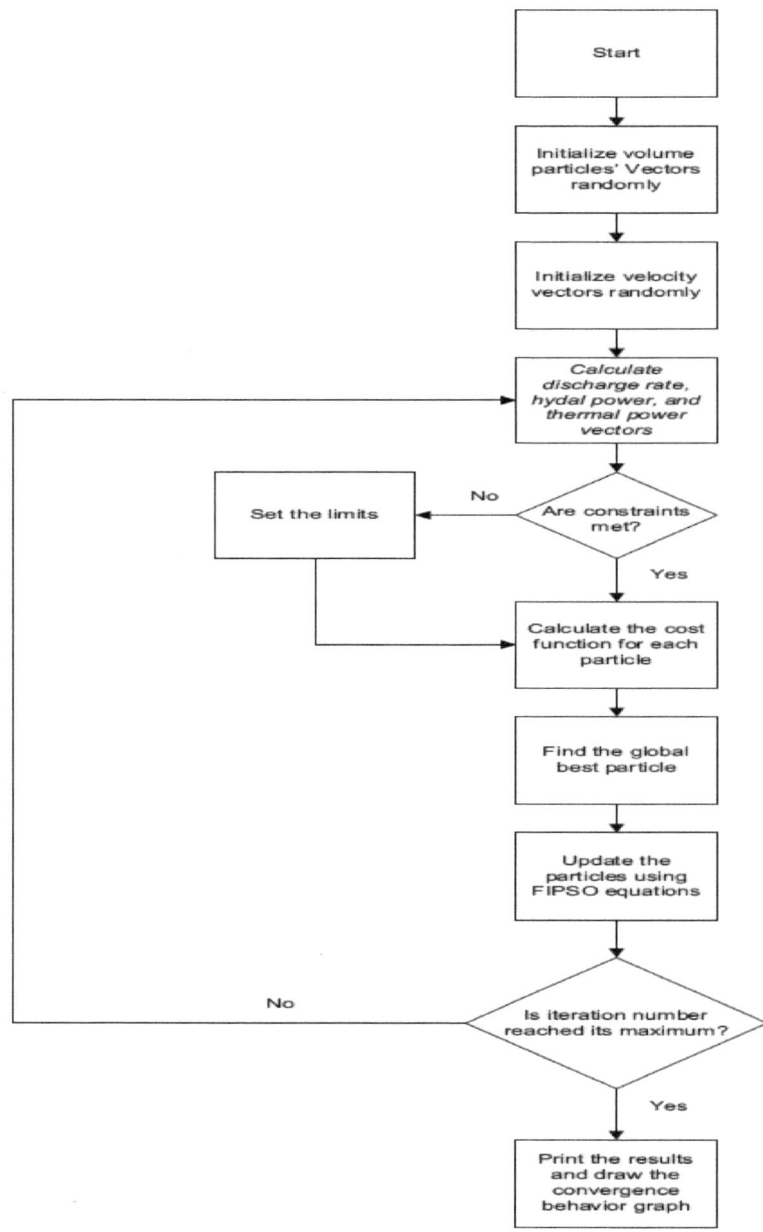

Figure 10: Flow chart of running FIPSO based solution [18].

Chapter 4

Results

4 Overview

In this chapter, the results of the simulations found in literature on the hydrothermal scheduling problem are presented. Moreover, a comparison is also presented among different variants of PSO. A comparison between all the works done on the very problem using different conventional and un-conventional optimization techniques is also provided. Also given are the results of the implementation of FIPSO algorithm on pumped storage hydrothermal scheduling problem.

4.1 Results of short term hydrothermal scheduling using PSO

4.1.1 Using volume of water as particle

The results of the research work by Samudai *et al* in [2] using Canonical PSO are shown in Table 3 and its convergence behavior is shown in Figure 11:

Table 3: Results of work by Samudai *et al* [2].

Interval	Thermal Power (MW)	Hydro Power (MW)	Volume of water (acre-ft.)	Discharge rate (Acre-ft/hr)	Total cost of operation ($)
1	812.5	387.4	96931.9	2255.6	
2	801.5	698.4	75318.3	3801.1	
3	1100.0	0	99318.3	0	
4	804.7	995.2	60000.0	5276.5	693428.5
5	950	0	84000.0	0	
6	561.5	738.4	60000.0	4000.0	

Its convergence behavior is shown next. It can be noticed that this algorithm does converge but since each particle is just influenced by the global best

31

particle of each of the iterations, the particle ultimately do stick to the local minima. The proposed algorithm will however, depict that it has a power to scan and span the complete search space.

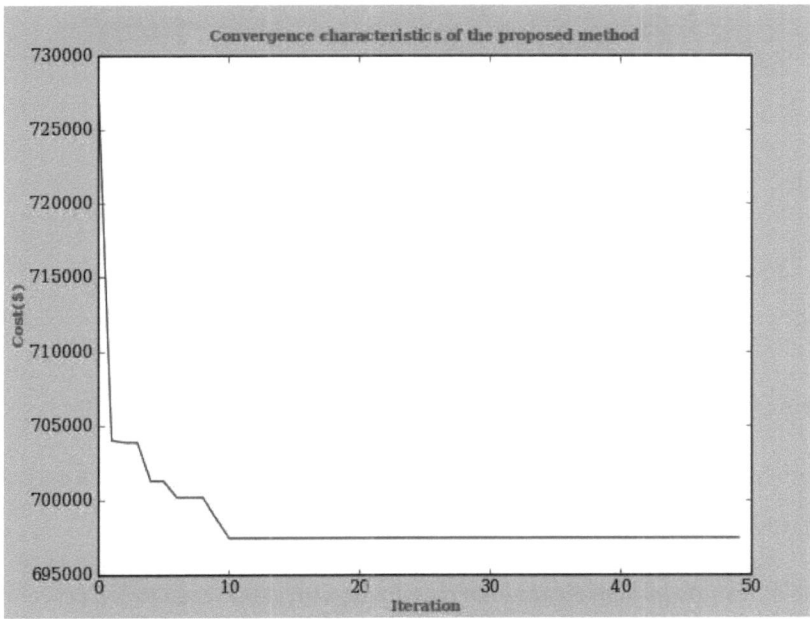

Figure 11: Convergence characteristics by work done by Samudai *et al* [2].

4.1.2 Using discharge rate as particles

When discharge rate was taken to be the independent variable and as a particle, the results attained are shown in Table 4.

Table 4: Results of short term hydrothermal scheduling using PSO when discharge rate is taken as particle [2].

Interval	Thermal Power (MW)	Hydro Power (MW)	Volume of water (acre-ft.)	Discharge rate (Acre-ft/hr)	Total cost of operation ($)
1	896.31	303.69	101928	1839.3	
2	896.31	603.69	85964	3330.3	
3	896.31	203.69	93857	1342.3	
4	896.31	903.69	60001	4821.3	709863.04
5	788.98	161.02	70439	1130.2	
6	788.98	511.02	60001	2869.8	

4.1.3 Using hydro power as particles

When hydro power was taken to be the independent variable and as a particle, the results attained were:

Table 5: Short term hydrothermal scheduling using PSO when hydro power is taken as particles [2].

Interval	Thermal Power (MW)	Hydro Power (MW)	Volume of water (acre-ft.)	Discharge rate (Acre-ft/hr)	Total cost of operation ($)
1	894.9	305.0	101846.7	1846.1	
2	900.3	599.6	86120.7	3310.4	
3	894.5	205.4	93908.1	1351.0	709862.7
4	895.4	904.5	600000	4825.6	
5	786.9	163.0	70314.5	1140.4	

| 6 | 791.0 | 503.9 | 60000 | 2859.5 | |

4.1.4 Short term hydrothermal scheduling using PSO taking thermal power as particles

When thermal power was taken to be the independent variable and as a particle, the results attained were:

Table 6: Short term hydrothermal scheduling using PSO taking thermal power as particles [2].

Interval	Thermal Power (MW)	Hydro Power (MW)	Volume of water (acre-ft.)	Discharge rate (Acre-ft/hr)	Total cost of operatior ($)
1	895.36	304.64	101871.3	1844.06	
2	897.97	602.02	86006.37	3322.07	
3	895.77	204.22	93866.26	1345.0	
4	896.15	903.84	600007	4822.12	709862.2
5	790.18	159.81	705095	1124.26	
6	787.76	512.23	600000	2875.79	

4.2 Results of short term hydrothermal scheduling using improved PSO

The results of the work by Padimini *et al* in [1], using Improved PSO are shown in Table 7.

Conclusion

Table 7: Result of the work by Padimini *et al* [1].

Interval	Thermal Power (MW)	Hydro Power (MW)	Volume of water (acre-ft.)	Discharge rate (Acre-ft/hr)	Total cost of operation ($)
1	812.54	387.45	96931.9	2255.6	
2	801.58	698.41	75318.3	3801.1	
3	1100.0	0	99314.3	0	
4	804.7	995.2	59996.0	5276.5	693426.2
5	950	0	83996.0	0	
6	561.5	738.43	59996.0	4000	

Its convergence rate is also shown in Figure 12.

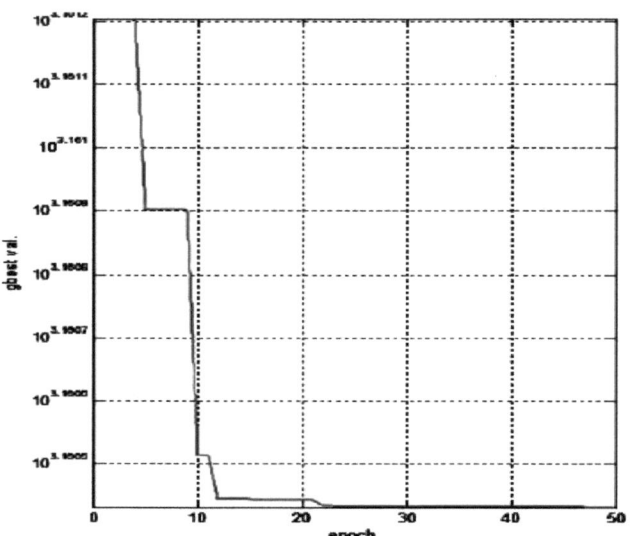

Figure 12: Convergence behavior of the work by Padimini *et al* [1].

35

4.3 Results of short term hydrothermal scheduling problem using hybrid technique

The result of short term hydrothermal scheduling problem using hybrid technique is shown in the form of Table 8 as given in reference [14].

Table 8: Results of short term hydrothermal scheduling using hybrid technique [14].

Interval	Thermal Power (MW)	Hydro Power (MW)	Volume of water (acre-ft.)	Total cost of operation ($)
1	912.9	287.0	102920	
2	883.4	616.5	86190	
3	876.2	223.7	95610	703180
4	893.7	906.2	61600	
5	777.2	172.7	71330	
6	773.9	526.0	60000	

4.4 Results of short term hydrothermal scheduling problem using simulated annealing technique

The simulated annealing optimization technique when applied on the very problem, the results were as shown in Table 9.

Table 9: Short term hydrothermal scheduling using simulated annealing with relaxed and non-relaxed constraints [14].

Interval	Thermal Power (MW)	Hydro Power (MW)	Volume of water (acre-ft.)	Discharge rate (Acre-ft/hr)	Total cost of operation ($)
1	893.7	306.2	101773.8	1852.1	
2	895.2	604.7	85746.0	3335.6	
3	884.3	215.6	92922.8	1401.9	
4	912.2	887.7	60015.6	4742.2	709874.3
5	781.8	168.1	70028.4	1165.6	
6	795.8	504.1	60000	2835.7	

Interval	Thermal Power (MW)	Hydro Power (MW)	Volume of water (acre-ft.)	Discharge rate (Acre-ft/hr)	Total cost of operation ($)
1	895.1	304.82	101860.2	1844.9	
2	894.0	605.96	85761.0	3341.6	
3	885.7	214.29	93020.8	1395.0	
4	911.7	888.24	60086.0	4744.5	709877.3
5	784.5	165.42	70260.1	1152.1	
6	791.95	508.05	60000.0	2855.0	

4.5 Results of short term hydrothermal scheduling using different meta heuristic techniques

Sinha *et al* in [16] performed the short term hydrothermal scheduling problem using different meta- heuristic optimization techniques, the results are shown in

Conclusion

Table 10 and Table 11.

It is however worth mentioning that in this research work, the convergence behavior of solution with respect to each of the iterations is not shown. It raises a question in mind about the convergence behavior of the algorithms implemented.

Table 10: Results of short term hydrothermal scheduling using genetic algorithm [16].

Interval	Thermal Power (MW)	Hydro Power (MW)	Volume of water (acre-ft.)	Discharge rate (Acre-ft/hr)	Total cost of operation ($)
1	896.86	301.14	101960.94	1836.59	
2	897.15	602.85	86046.83	3326.18	
3	893.85	206.15	93791.98	1354.57	
4	897.38	902.62	60000	4816.01	709863.56
5	794.45	155.55	70763.09	1103.08	
6	783.52	516.48	60000.01	2896.92	

Table 11: Results of short term hydrothermal scheduling using PSO [16].

Interval	Thermal Power (MW)	Hydro Power (MW)	Volume of water (acre-ft.)	Discharge rate (Acre-ft/hr)	Total cost of operation ($)
1	896.31	303.69	101928.40	1839.30	
2	896.31	603.69	85964.80	3330.30	709862.048
3	896.31	203.69	93857.20	1342.30	

38

4	896.31	903.69	60001.60	4821.30	
5	788.98	161.02	70439.20	1130.20	
6	788.98	511.02	60001.60	2869.80	

4.6 Comparison among the best implementations of PSO variants

As it was discussed in chapter 4, the two forms of PSO had been implemented earlier on the same short term hydro-thermal scheduling problem. One by Samudai *et al* and the other by Padamini *et al*. [1, 2].

Table 12: Comparison among Two best PSO implementations so far.

Works	Minimum cost ($)
Samudai*et al* [2]	693428.5
Padamini*et al*[1]	693426.2

The results are obviously depicting that the PSO variants have produced better costs for the short term hydrothermal scheduling problem.

4.7 Comparison among all the implementations of PSO on short term hydrothermal scheduling problem

Short term hydro-thermal scheduling has been a very famous problem since 1984, different works have been done and different algorithms have been proposed to find a better result to this problem. Table 19 gives this comparison. It is quite astonishingly interesting that PSO variants have produced better results as compared to other algorithms.

Table 13: Comparison of result of proposed method with results of previously done works.

Number	Researcher	Algorithm	Minimum cost ($)
1	Alen J. Wood[4]	Gradient Search	709877.38
2	Sinha *et al*[17]	Fast evolutionary programming	709862.05
3	Wong K. Patel[15]	Simulated annealing	709874.36
4	Sinha *et al*[16]	GAF	709863.70
5	Sinha *et al*[16]	CEP	709862.65
6	Sinha *et al*[16]	FEP	709864.59
7	Sinha *et al*[16]	Particle Swarm Optimization	709862.048
8	D.S. Suman*et al*[14]	Hybrid Evolutionary Programing	703180.26
9	Samudi*et al*[2]	Particle Swarm Optimization	696002.3
10	Padamini*et al*[1]	Improved PSO Optimization	693426.2

4.8 Results of Pumped Storage Short Term Hydrothermal Scheduling Problem Using FIPSO

It is worth mentioning that reference [4] has implemented gradient search algorithm on the pumped storage hydrothermal scheduling problem and has found the global best result. FIPSO will also reach those results and the convergence behavior for accelerated PSO, another variant of PSO is given along with the results in Figure 13 and Table 14 respectively.

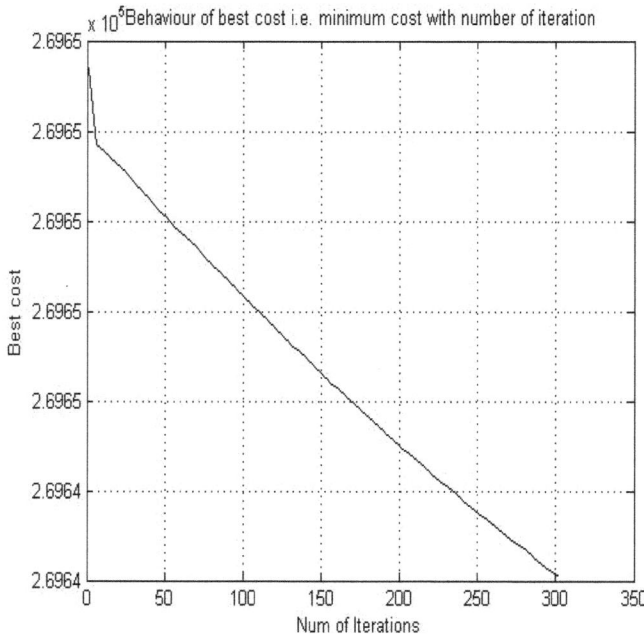

Figure 13: Convergence behavior of best cost with number of iterations in pumped storage hydrothermal scheduling problem using accelerated PSO.

Table 14: Results of final iteration of pumped storage hydrothermal scheduling using PSO variants.

Interval	Thermal Power (MW)	Hydro Power (MW)	Volume of water (acre-ft.)	Discharge rate (Acre-ft/hr)	Total cost of operation ($)
1	1450	150	60000	500	
2	1500	300	28000	800	
3	1450	150	800	500	
4	800	-300	3200	-600	269642.4
5	800	-300	5600	-600	
6	800	-300	8000	-600	

Conclusion

Hydrothermal scheduling, which is a renowned optimization problem even till today, can be solved with variants of PSO algorithm to find the minimum scheduling cost. However, due to the meta-heuristic nature of these algorithm, the results are not always approaching the near global optimal solution. Different neighborhood topolog es can be utilized to improve this problem.

Future Scope

Particle Swarm Optimization and its variants, especially FIPSO and others which have the potential to converge towards the global optimum solution of different multi-modal and non-linear optimization problems have a great scope in the area of power systems. There are several optimization problems which require the solution that may lead to the minimum cost of operation. Such problems are economic dispatch problem, unit commitment problem, optimal load shedding problem, hydrothermal scheduling problem, and optimal power flow problem, take or pay fuel agreement problem and many others. A very famous transmission systems problem known as VAR compensation requires the optimal operation of compensation devices, mostly FACTS devices these days, can also be implemented using FIPSO and variants like accelerated PSO.

Many power electronics applications can be performed using FIPSO and other variants of PSO. The maximum power point trackers may also operate using these algorithms.

In short, all those problems which can be modeled in a mathematical form and an objective function is obtained along with the constraints of optimality can be solved using these techniques. In countries like Pakistan, where load shedding is a big issue and country is always in one or another crisis, the savings in operations of every system can be made possible using these sorts of algorithms. So, it will not be a big thing to say that meta-heuristic optimization techniques is a grey area of research in which many more findings will be made possible in future.

References

[1]   S. Padmini and C. Rajan, "Improved PSO for short term hydrothermal scheduling", *Proceedings of International Conference on Sustainable Energy and Intelligent Systems, (SEISCON 2011)*, 2011, pp. 332-334.

[2]   C. Samudi, G. Das, P. Ojha, T. Sreeni, and S. Cherian, "Hydro thermal scheduling using particle swarm optimization", *Proceedings of Transmission and Distribution Conference and Exposition, 2008. T&# x00026; D. IEEE/PES*, pp. 1-5.

[3]   S. Sivanagaraju, *Power system operation and control*: Pearson Education India, 2009.

[4]   A. J. Wood and B. F. Wollenberg, *Power generation, operation, and control*: John Wiley & Sons, 2012.

[5]   D. Fouskakis and D. Draper, "Stochastic optimization: a review,"*Proceedings of International Statistical Review"*, 2002, vol. 70, pp. 315-349.

[6]   W. Chang, "Optimal Scheduling of Hydrothermal System Based on Improved Particle Swarm Optimization", *Proceedings of Power and Energy Engineering Conference (APPEEC), Asia-Pacific*, 2010, pp. 1-4.

[7]   H. B. Tavakoli, B. Mozafari, and S. Soleymani, "Short-Term Hydrothermal Scheduling via Honey-Bee Mating Optimization Algorithm", *Proceedings of Power and Energy Engineering Conference (APPEEC), Asia-Pacific*, 2012, pp. 1-5.

[8]   S. Thakur, C. Boonchay, and W. Ongsakul, "Optimal hydrothermal generation scheduling using self-organizing hierarchical PSO", *Proceedings of Power and Energy Society General Meeting, IEEE*, 2010, pp. 1-6.

[9]   R. Mendes, J. Kennedy, and J. Neves, "The fully informed particle swarm:

simpler, maybe better" *Evolutionary Computation, IEEE Transactions on,* vol. 8, pp. 204-210, 2004.

[10] X.-S. Yang, *Engineering optimization: an introduction with metaheuristic applications*, John Wiley & Sons, 2010.

[11] J. Kennedy and R. Mendes, "Neighborhood topologies in fully informed and best-of-neighborhood particle swarms",*IEEE Transactions on Systems Man and Cybernetics Part C Applications and Reviews,* vol. 36, p. 515, 2006.

[12] J. Kennedy and R. Mendes, "Population structure and particle swarm performance", 2002.

[13] A. J. R. Medina, G. T. Pulido, and J. G. Ramírez-Torres, "A Comparative Study of Neighborhood Topologies for Particle Swarmss Optimizers", in *IJCCI*, 2009, pp. 152-159.

[14] C. Nallasivan, D. Suman, J. Henry, and S. Ravichandran, "A novel approach for short-term hydrothermal scheduling using hybrid technique", Proceedings of *Power India Conference, IEEE*, 2006, p. 5.

[15] K. Wong and Y. Wong, "Short-term hydrothermal scheduling part. I. Simulated annealing approach",*IEE Proceedings-Generation, Transmission and Distribution,* 1994,vol. 141, pp. 497-501.

[16] N. Sinha and L.-L. Lai, "Meta heuristic search algorithms for short-term hydrothermal scheduling",*Proceedings of International Conference on Machine Learning and Cybernetics,* 2006, pp. 4050-4056.

[17] N. Sinha, R. Chakrabarti, and P. Chattopadhyay, "Fast evolutionary programming techniques for short-term hydrothermal scheduling",*Electric Power Systems Research,* vol. 66, pp. 97-103, 2003.

[18] M.S. Fakhar, S.A.R. Kashif, M.A. Saqib and T. ul Hassan, "Non cascaded short-term hydro-thermal scheduling using fully-informed particle swarm optimization". *International Journal of Electrical Power & Energy*

Systems, 2015, vol. *73*, pp.983-990.

[19] M.S. Fakhar, S.A.R. Kashif and M.A. Saqib, "particle swarm optimization and its variants for short term hydrothermal scheduling". *Science International Lahore, 2014,* vol. 26 (4), pp 1489-1494.

[20] M.S. Fakhar, S.A.R. Kashif, H.Z. Hussain and B.A. Ahmad, "implementation of PSO and FIPSO with consideration of constant and linearly decreasing weight strategies on Michaelwicz 3-D function". *Science International Lahore, 2015,* vol 27 (5), pp 4097-4100.

Appendices

A1: MATLAB program to solve the 3-D Michaelewicz function using canonical PSO

The MATLAB program to solve the 3-D Michaelewicz function using canonical PSO algorithm is;

```
% write a function named canonical_pso_demo that takes number of particles
"n"
% and Number of iterations as inputs and outputs the 3 column table with
name "best"

function [best] = canonical_pso_demo (n,Num_iterations)

ifnargin<2
Num_iterations = 20;
end

ifnargin< 1
    n = 20;
end


fstr = '-sin (x) * (sin (x^2/pi)) ^ 20 - sin (y) * (sin (2*y^2/pi))^20';


f = vectorize (inline (fstr));


range = [0 4 0 4];
```

```
vrange = [-0.2667 0.2667 -0.2667 0.2667];
alpha = 0.2;
beta = 0.5;


Ndiv = 100;
dx = (range (2) - range (1))/Ndiv
dy = (range (4) - range (3))/Ndiv
xgrid = range (1):dx:range (2);
ygrid = range (3):dy:range (4);


[x,y] = meshgrid (xgrid,ygrid);


z = f (x,y);


figure (1);
surfc (x,y,z);


best = zeros (Num_iterations,3)


% initiallize local best arrays and present valuse of x and y


[xn,yn,pbx,pby] = init_pso (n,range);


% initiallize velocity also


[vxn,vyn] = init_vel (n,Num_iterations,range);
```

% till this point, we have initialized and thus found the curentrandome

% positions of particles, current local bests of all particles and current

% velocity for all particles.

figure (2);

% here now we need to upgrade our current particles and get the result

% itratively till the maximum number of iterations.


```
fori = 1:Num_iterations
contour (x,y,z,20);
hold on;
   % xo,yo,zo are the global bests of each iterations
zn = f (xn,yn); % here we have first calculated our fitness function.
zpb = f (pbx,pby);
zn_min = min (zn);


   % here we are initiallizing our global bests.


xo = min (xn(zn==zn_min));
yo = min (yn(zn==zn_min));
zo = min (zn(zn==zn_min));


plot (xn,yn,'.',xo,yo,'*');
axis (range);


   % here we are now going to move our current positions of xn and yn to
```

Appendices

% new positions of xn and yn.a

```
[xn,yn] = pso_move (xn,yn,zn,zpb, xo,yo,vxn,vyn,pbx,pby,range,vˆange);

drawnow;

hold off;

best (i,1) = xo;
best (i,2) = yo;
best (i,3) = zo;
end

function [xn,yn,pbx,pby] = init_pso (n,range)

xrange = range (2) - range (1);
yrange = range (4) - range (3);

xn = rand (1,n)*xrange+range(1);
yn = rand (1,n)*yrange+range(3);

pbx = xn;
pby = yn;
```

% write function for velocity

```
function [vxn,vyn] = init_vel (n,Num_iterations,range)
Num_iterations = Num_iterations;
xmax = range (2);
xmin = range (1);
ymax = range (4);
ymin = range (3);


vxmax = (xmax - xmin)/Num_iterations;
vymax = (ymax - ymin)/Num_iterations;


vxn = -vxmax + (vxmax + vxmax).* rand (1,n);
vyn = -vymax + (vymax + vymax).* rand (1,n);



function [xn,yn] = pso_move (xn,yn,zn,zpb,
xo,yo,vxn,vyn,pbx,pby,range,vrange)
  % local best array for x and is equal to that calculated in the previous iteration
  % local best array for y and is equal to that calculated in the previous iteration


nn = size (yn,2);


% now i just need to upgrade the local best arrays for xn and yn and then
% need to upgrade vxn and vyn
fori = 1:nn
if  zn (i) <zpb (i)


pbx (i) = xn (i);
```

```
pby (i) = yn (i);
end
end



pbx = pbx;
pby = pby;

  % now update the velocity

vxn = vxn + (2.5 * rand (1) * (pbx - xn) + (2.5 * rand (1) * (xo - xn)));
vyn = vyn + (2.5 * rand (1) * (pby - yn) + (2.5 * rand (1) * (yo - yn)));

  [vxn,vyn] = findrange_v (vxn,vyn,vrange);

% now findxn, ynetc
% nn = size (yn,2);

xn = xn + vxn;
yn = yn + vyn;
zn = zn;
zpb = zpb;

[xn,yn] = findrange(xn,yn,range);

function [xn,yn] = findrange (xn,yn,range)
```

```
nn = length (yn);
fori =1:nn
ifxn(i) <= range (1)
xn(i) = range (1);
end
ifxn(i) >= range (2)
xn(i) = range (2);
end
ifyn(i) <= range (3)
yn(i) = range (3);
end
ifyn(i) >= range (4)
yn(i) = range (4);
end
end


function [vxn,vyn] = findrange_v (vxn,vyn,vrange)
nn = length (vyn);
fori =1:nn
ifvxn(i) <= vrange (1)
vxn(i) = vrange (1);
end
ifvxn(i) >= vrange (2)
vxn(i) = vrange (2);
end
ifvyn(i) <= vrange (3)
```

```
vyn(i) = vrange (3);
end
ifvyn(i) >= vrange (4)
vyn(i) = vrange (4);
end
end
```

A 2: MATLAB program to solve the 3-D Michaelewicz function using FIPSO

MATLAB program to solve the 3-D Michaelewicz function using FIPSO is alsc given for the reader to understand how FIPSO program works.

```
% write a function named FIPSO_pso_demo that takes number of particles "n"
% and Number of iterations as inputs and outputs the 3 column table with
name "best"

function [best] = FIPSO_pso_demo (n,Num_iterations)

ifnargin<2
Num_iterations = 20;
end

ifnargin< 1
  n = 20;
end
```

```
fstr = '-sin (x) * (sin (x^2/pi)) ^ 20 - sin (y) * (sin (2*y^2/pi))^20';

f = vectorize (inline (fstr));

range = [0 4 0 4];
vrange = [-0.2667 0.2667 -0.2667 0.2667];
alpha = 0.2;
beta = 0.5;

Ndiv = 100;
dx = (range (2) - range (1))/Ndiv;
dy = (range (4) - range (3))/Ndiv;
xgrid = range (1):dx:range (2);
ygrid = range (3):dy:range (4);

[x,y] = meshgrid (xgrid,ygrid);

z = f (x,y);

figure (1);
surfc (x,y,z);

best = zeros (Num_iterations,3);

% initiallize local best arrays and present valuse of x and y

[xn,yn,pbx,pby] = init_pso (n,range);
```

```
% initiallize velocity also

[vxn,vyn] = init_vel (n,Num_iterations,range);

% initiallizevel_sum_x

vel_sum_x = rand (1,n);

% initiallizevel_sum_x

vel_sum_y = rand (1,n);

% till this point, we have initialized and thus found the current random
% positions of particles, current local bests of all particles and current
% velocity for all particles.

figure (2);

% here now we need to upgrade our current particles and get the result
% iteratively till the maximum number of iterations.

fori = 1:Num_iterations
contour (x,y,z,20);
hold on;

    % xo,yo,zo are the global bests of each iterations
```

```
zn = f (xn,yn); % here we have first calculated our fitness function.
zpb = f (pbx,pby);
zn_min = min (zn);


    % here we are initiallizing our global bests.


xo = min (xn(zn==zn_min));
yo = min (yn(zn==zn_min));
zo = min (zn(zn==zn_min));


plot (xn,yn,'.',xo,yo,'*');
axis (range);


    % here we are now going to move our current positions of xn and yn to
% new positions of xn and yn.


    [xn,yn] = pso_move (xn,yn,zn,zpb,vel_sum_x,vel_sum_y,
    xo,yo,vxn,vyn,pbx,pby,range,vrange);
drawnow;
hold off;


best (i,1) = xo;
best (i,2) = yo;
best (i,3) = zo;
end
```

```
function [xn,yn,pbx,pby] = init_pso (n,range)


xrange = range (2) - range (1);
yrange = range (4) - range (3);


xn = rand (1,n)*xrange+range(1);
yn = rand (1,n)*yrange+range(3);


pbx = rand (1,n) * xrange + range (1);
pby = rand (1,n) * yrange + range (3);


% write function for velocity


function [vxn,vyn] = init_vel (n,Num_iterations,range)
Num_iterations = Num_iterations;
xmax = range (2);
xmin = range (1);
ymax = range (4);
ymin = range (3);


vxmax = (xmax - xmin)/Num_iterations;
vymax = (ymax - ymin)/Num_iterations;


vxn = -vxmax + (vxmax + vxmax).* rand (1,n);
vyn = -vymax + (vymax + vymax).* rand (1,n);
```

```
function [xn,yn] = pso_move (xn,yn,zn,zpb,vel_sum_x,vel_sum_y,
xo,yo,vxn,vyn,pbx,pby,range,vrange)


    % local best array for x and is equal to that calculated in the previous
    % iteration
   % local best array for y and is equal to that calculated in the previous
    % iteration


nn = size (yn,2);


% now i just need to upgrade the local best arrays for xn and yn and then
% need to upgrade vxn and vyn


fori = 1:nn
if  zn (i) <zpb (i)


pbx (i) = xn (i);
pby (i) = yn (i);
end
end




pbx = pbx;
pby = pby;


    % now i need to update the velocity
```

```
% first calculate vel_sum_x and vel_sum_y


fori = 1:nn
for j = 1:nn
vel_sum_x (i) = vel_sum_x (i) + (4.1 * rand (1) * (pbx (j) - xn (i)));
end
end


vel_sum_x = vel_sum_x/nn;
vxn = 0.729*(vxn + vel_sum_x);


fori = 1:nn
for j = 1:nn
vel_sum_y (i) = vel_sum_y (i) + (4.1 * rand (1) * (pby (j) - yn (i)));
end
end


vel_sum_y = vel_sum_y/nn;
vyn = 0.729* (vyn + vel_sum_y);



%vxn = vxn + (2.5 * rand (1) * (pbx - xn) + (2.5 * rand (1) * (xo - xn)));
%vyn = vyn + (2.5 * rand (1) * (pby - yn) + (2.5 * rand (1) * (yo - yn)));


[vxn,vyn] = findrange_v (vxn,vyn, vrange);


% now findxn, ynetc
```

```
% nn = size (yn,2);


xn = xn + vxn;
yn = yn + vyn;


%zn = zn;
%zpb = zpb;


[xn,yn] = findrange(xn,yn,range);


function [xn,yn] = findrange (xn,yn,range)
nn = length (yn);
fori =1:nn
ifxn(i) <= range (1)
xn(i) = range (1);
end
ifxn(i) >= range (2)
xn(i) = range (2);
end
ifyn(i) <= range (3)
yn(i) = range (3);
end
ifyn(i) >= range (4)
yn(i) = range (4);
end
end
```

```
function [vxn,vyn] = findrange_v (vxn,vyn,vrange)
nn = length (vyn);
fori =1:nn
ifvxn(i) <= vrange (1)
vxn(i) = vrange (1);
end
ifvxn(i) >= vrange (2)
vxn(i) = vrange (2);
end
ifvyn(i) <= vrange (3)
vyn(i) = vrange (3);
end
ifvyn(i) >= vrange (4)
vyn(i) = vrange (4);
end
end
```

yes

I want morebooks!

Buy your books fast and straightforward online - at one of world's fastest growing online book stores! Environmentally sound due to Print-on-Demand technologies.

Buy your books online at
www.morebooks.shop

Kaufen Sie Ihre Bücher schnell und unkompliziert online – auf einer der am schnellsten wachsenden Buchhandelsplattformen weltweit! Dank Print-On-Demand umwelt- und ressourcenschonend produziert.

Bücher schneller online kaufen
www.morebooks.shop

info@omniscriptum.com
www.omniscriptum.com

MIX
Papier aus verantwortungsvollen Quellen
Paper from responsible sources
FSC® C105338

Printed by Books on Demand GmbH, Norderstedt / Germany